D1783978

Trilobites in Wales

R.M. Owens

Department of Geology, National Museum of Wales, Cardiff

AMGUEDDFA GENEDLAETHOL CYMRU
NATIONAL MUSEUM OF WALES

Geological Series No. 7

CARDIFF, APRIL 1984

© National Museum of Wales 1984
ISBN 0 7200 0289 3

Designed by Penknife, Cardiff
Typeset by Afal, Cardiff
Printed by South Western Printers, Caerphilly

Cover illustration
Hamatolenus (Myopsolenus) douglasi Bassett, Owens & Rushton; from the Lower Cambrian
Hell's Mouth Grits, E side of Hell's Mouth (Porth Neigwl,) St Tudwal's Peninsula, Gwynedd;
x8.5 approximately.

Note
This booklet is a revised and expanded version of an article published under the title *In search of
Welsh Trilobites* in *Amgueddfa, Bulletin of the National Museum of Wales,* No.9, Winter 1971, pp.24-37.

Trilobites in Wales

'A good collection of well-arranged trilobites looks better in the cabinet than perhaps any other fossils', wrote J.E. Taylor in *Our Common British Fossils*, published in 1885. Such is their popular appeal that trilobites have always been some of the most eagerly sought after of all fossils. Their name, suggested by their singular three-lobed appearance is derived from 'Trilobitae', introduced by the German naturalist Johann Walch in 1771 in his *Der Naturgeschichte der Versteinerungen* ('Natural History of Petrefactions'). The study of trilobites has particularly long associations with Wales, and the ancient rocks which crop out over much of the Principality have been well known as a rich source of them for nearly 300 years. This article outlines some of the history of their investigation in the area, describes their occurrence there, and discusses aspects of their nomenclature and morphology which are well illustrated by Welsh examples.

Edward Lhuyd, 'flat-fish' and Trinucleum

The 17th Century saw the beginnings of the great period of collecting in natural history, and interest in acquiring all kinds of zoological, botanical and geological specimens increased steadily in the 18th and early 19th Centuries, and then much more rapidly during the Victorian era. Whilst many people collected for acquisitive reasons alone, the early naturalists began to prepare detailed monographs, often copiously illustrated with beautifully executed woodcuts and lithographs. The nature of fossils was still a matter of considerable debate in the 17th and early 18th Centuries, when they were often referred to as 'formed' or 'figured' stones. Although few naturalists at that time could explain satisfactorily what they were, several theories were put forward, including origins through supernatural forces, and whilst fossils were commonly compared with living organisms, they were rarely believed to have been originally organic themselves.

The first descriptions and illustrations of trilobites were made by Edward Lhuyd (1660-1709), the famous 17th Century Welsh naturalist. He became an assistant at the Ashmolean Museum, Oxford when it opened in 1683, and worked under Robert Plot, the first Keeper, whom he succeeded in 1691. He travelled extensively throughout the Welsh countryside gathering information for his intended *Natural History of Wales* for which he drew up 'a design' in 1695, but which unfortunately was never finished. Whilst on his travels Lhuyd wrote letters (many of which were published in the

Philosophical Transactions of the Royal Society) to contemporary scientists in Britain and abroad. In one of these, written to Professor Rivinum in Leipzig from Caldy Island in March 1698, he illustrated trilobites from the Llandeilo district, Carmarthenshire (Dyfed), which he described as *Trinucleum* and *Buglossam curtam strigosam* or *flat-fish*. In August 1698 he wrote to Dr. Martin Lister: 'I should have troubled you with some sort of Account of our Travels; which, as you'll find by the inclosed Draughts of Figured Stones, has been tolerably successful. The 8, 9 and 15th we found near the Llan Deilo (Llandeilo) in Carmarthenshire; . . . The 15th whereof we found great plenty must doubtless be referred to the sceleton of some Flat-Fish; the 8th and 9th I know not at all what to make of'. In the following year Lhuyd published the first ever catalogue of fossils, written in Latin and entitled *Lithophylacii Britannici Ichnographia*, in which were reproduced several of the trilobite illustrations included in his letters. Lhuyd's illustrations are sufficiently accurate to enable determination of the trilobites. His *flat-fish* are now known as *Ogygiocarella debuchii*. The resemblance of this species to the skeleton of a large flat fish is indeed an apt and striking one when it is considered that no animal closely resembling this fossil is known today. *Trinucleum* literally means 'three nuts' – reference to the three smooth lobes on the headshield. In terms of modern classification it probably belongs either to *Lloydolithus* or to *Marrolithus* (the name *Trinucleus* is now restricted to a form which occurs abundantly near Llandrindod Wells, and to applied species). The *flat-fish* and *Trinucleum* were based upon complete, or nearly complete specimens. Two specimens figured by Lhuyd are incomplete – parts of headshields only – and it is these which completely mystified him (see above). One belongs probably to *Marrolithus favus,* the other to *Atractopyge verrucosa*. A specimen of the latter, which is almost certainly Lhuyd's original, was identified some years ago in the collections of the University Museum, Oxford, by the former Curator, Mr. J.M. Edmonds. Lhuyd illustrated both specimens upside down, and as mirror images of the originals (as in many early woodcuts).

A further *flat-fish* was illustrated by James Parkinson in Volume Three of his *Organic Remains of a Former World*, published in 1811. He commented as follows: 'Another species of this animal is found in the schistose strata in the neighbourhood of Llanelly (presumably Llandeilo is meant), in Carmarthenshire. . . .the outline of the animal approaches much nearer to the elliptical than the ovate form. From this latter circumstance, it

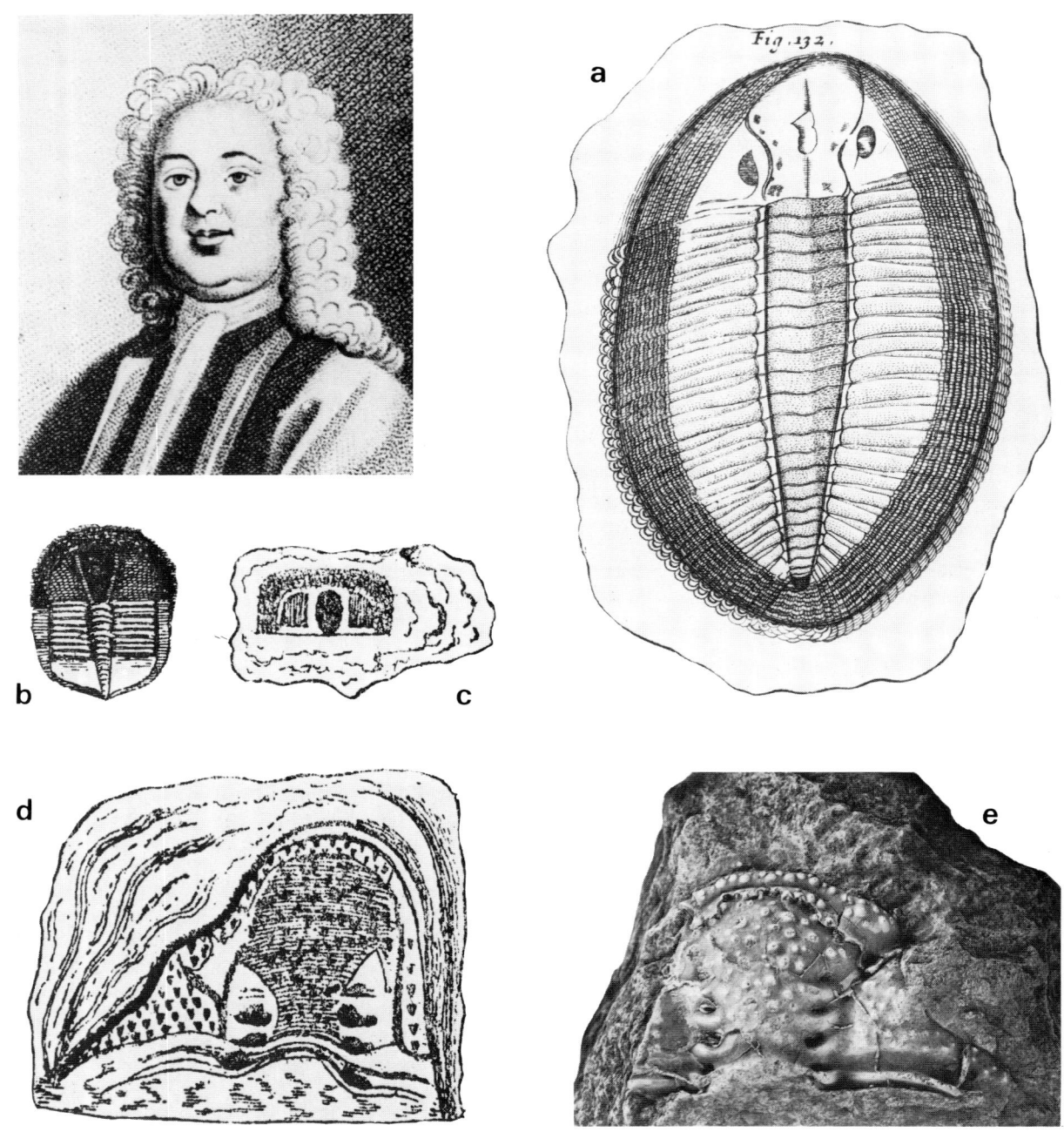

Edward Lhuyd, and his trilobites from Wales: a, *'Flat-fish'*; b, *'Trinucleum'*; c, *the 9th and* d, *the 8th 'Figured Stones' of his letter to Martin Lister;* e, *part of headshield of* Atractopyge verrucosa, *almost certainly the original of d (photograph courtesy British Geological Survey, London).*

obtains some slight resemblance to a sole, and has therefore been considered by some as the petrifaction of a fish of that tribe. The mutilated remains of this species, in consequence of the fossil being frequently severed transversely, have been regarded as petrified butterflies. . . On the remains of one of these I have perceived a very curious structure: it is in that part of the fossil which presents itself to view on the removal of the external covering, and which was probably the cuticle of the animal. Here the form of the parts appears exactly to correspond with that of the crustaceous covering, being transversely and somewhat obliquely disposed; but, aided by the lens, the eye discovers, that this pellicle is marked by frequent and regular rugae . . .'. This appears to be one of the earliest references to detailed morphology of the dorsal exoskeleton in trilobites. It also brings in a reference to petrified butterflies with which trilobite tails have been commonly associated.

Trilobites similar to Lhuyd's *flat-fish* gave rise to a local legend in the Carmarthen area, involving the Arthurian magician Merlin (from whom Carmarthen derives its name), as outlined by W.S. Symonds (1872) in his *Record of the Rocks*: 'An old legend also connects the fossils of Pensarn, and Mount

Pleasant, with the deeds of the great magician, whose last days were as singular as the earlier portion of his life. He fell in love with an angel sprite, or fair fay, without succeeding however in gaining her affection in return. One summer's day when the birds were singing, and the butterflies flitting, the wizard and the fairy entered a rocky cave, and here by the aid of a spell taught her by Merlin himself, the fairy closed the cavern and entombed the magician and the butterflies. Thus Merlin was "lost to life, and use, and name, and fame," and hence the appearance of the butterflies (or trilobites' tails) in the rocks of Mount Pleasant.' The trilobites in question have a broad resemblance to *O. debuchii*, but have recently been shown to belong to a new genus which has been named *Merlinia* from its association with this legend and from its common occurrence in the Carmarthen area.

The beginnings of modern studies
By the end of the 18th Century, it had become generally accepted that fossils were the remains of once living organisms, and through the pioneer work of the English engineer William Smith it became realized in the early 19th Century that fossils occur

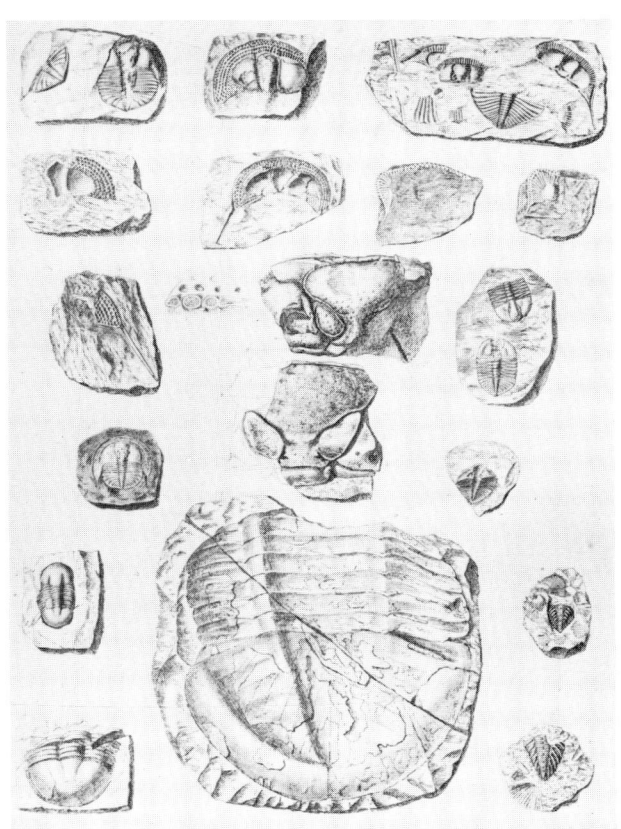

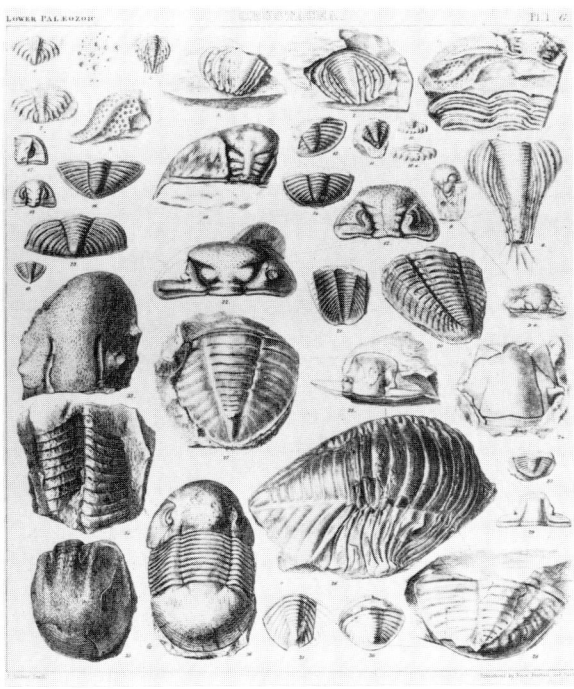

(left) *Plate 23 of Murchison's Silurian System (1839), showing trilobites from the Llandeilo and Caradoc series.* (above) *Plate 1G of Sedgwick and M'Coy's British Palaeozoic Rocks and Fossils (1851), showing Ordovician trilobites from North Wales and Cumbria. The figures were engraved by J.W. Salter.*

through rock strata in a regular order. The two names most intimately connected with establishing the sequence of the ancient rocks of Wales and the fossils occurring in them are those of Adam Sedgwick, Professor of Geology at the University of Cambridge, and Sir Roderick Murchison, a distinguished amateur, who was to become Director General of the Geological Survey of Great Britain. In the 1830s Sedgwick and Murchison began to investigate the geology of north and south Wales, respectively. The results of Sedgwick's work appeared mostly in learned journals, and the large numbers of fossils which he and his colleagues collected were added to the collections of the Sedgwick (then the Woodwardian) Museum, Cambridge. Some of these were described in illustrated catalogues, such as those published in conjunction with F. McCoy and J.W. Salter, and included many fine trilobites from Wales. Murchison's researches were published in a splendid and copiously illustrated memoir, published in 1839 and entitled *The Silurian System, founded on geological researches in the counties of Salop, Hereford, Radnor, Montgomery, Caermarthen, Brecon, Pembroke, Monmouth, Gloucester, Worcester and Stafford.* Of the 31 plates containing illustrations of the more important fossils, six are devoted to trilobites. It was the work of Sedgwick, Murchison and their colleagues which first brought many of the now famous Welsh trilobite localities to attention; their illustrations include new ones of Lhuyd's *Trinucleum* and *flat-fish*, besides a wealth of other kinds. Murchison even proposed naming one locality near Welshpool *Trilobite Dingle:* 'In a woody dingle . . . the shale abounds with beautifully ornamented trilobites of the genus *Trinucleus* . . .' (and in footnote): 'As this ravine has not, as far as I could ascertain, any name, I venture to hope that, to mark so interesting a fossil locality, Lord Clive (owner of Powis Castle, in whose grounds the dingle is situated) will call it 'Trilobite Dingle'. The large specimen, *Asaphus Powisii* (a trilobite), named in honour of the noble family . . . was found at this spot'. The name 'Trilobite Dingle' still survives as an informal name among geologists. Many specimens from this locality found their way to the local Powisland Museum at Welshpool, and in 1962 the material was deposited in the National Museum of Wales on permanent loan.

Along with the commencement of Murchison's and Sedgwick's work, the 1830s saw the establishment of the Geological Survey of Great Britain as a branch of the Ordnance Survey. As officers began to map the country on a systematic scale, fossils, trilobites among them, were collected

J.W. Salter (photograph courtesy of British Geological Survey, London, N.E.R.C. copyright).

as a means of dating and correlating the rocks. For the purpose of investigating these fossils so that they could be used to maximum advantage, the Survey appointed palaeontolgists. Of these, John William Salter was the one most intimately connected with much of the early research on British trilobites. Salter first gained an interest and knowledge of fossils through assisting with illustrations for several major works, including Murchison's *Silurian System.* In 1842 he worked under Sedgwick in the Woodwardian Museum at Cambridge helping to arrange fossil collections, and during the following three years undertook fieldwork in Wales under Sedgwick's expert guidance. In this way Salter developed a keen interest in Welsh palaeontology in general and in trilobites in particular, and in 1846 became assistant to Professor Edward Forbes, Palaeontologist at the Geological Survey; in 1854 he succeeded Forbes in this post, which he held until 1863. Salter's great interest in trilobites often carried him to Wales, and he made several exciting discoveries, perhaps the most noteworthy of which was that of an enormous trilobite, nearly two feet

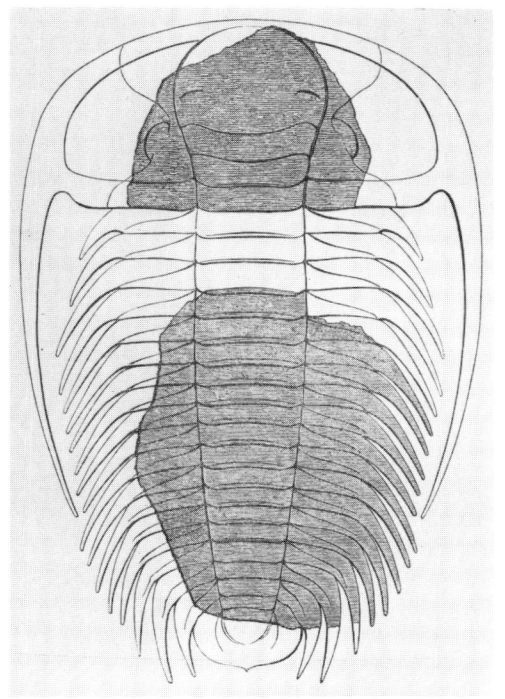

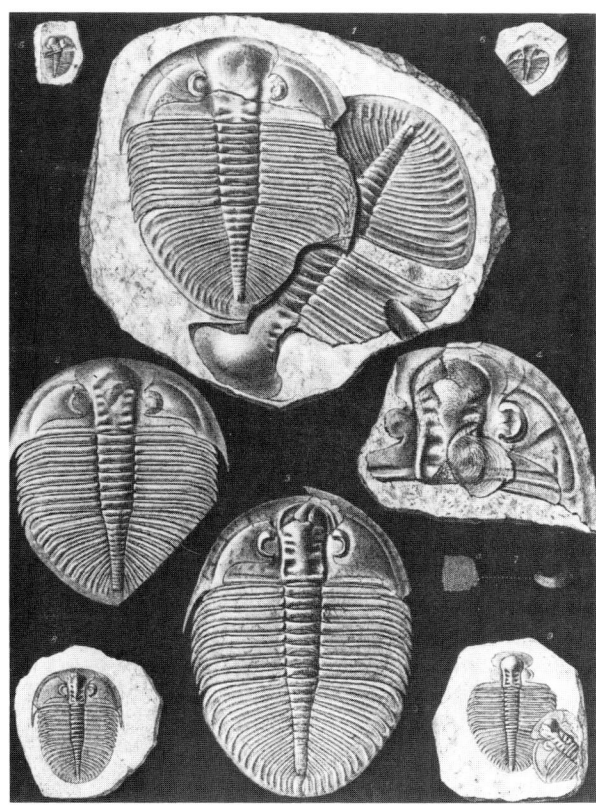

(left) *Salter's original reconstruction of the giant trilobite* Paradoxides davidis, *from Porth-y-Rhaw.* **(right)** *A plate from Salter's* Monograph of British Trilobites, *illustrating* Ogygiocarella debuchii *from the Builth Wells area.*

long, at Porth-y-Rhaw, near St. David's in 1862. He informed the Fellows of the Geological Society of London of his discovery in a paper read before them in February 1863 in the following words: 'My object now is to point out the locality and geological place of a giant Trilobite long looked for in Britain, and lately, I must say accidentally, found by me. I believed I was working at Solva Harbour, in Llandeilo Flags, but by good fortune I had landed instead in a parallel creek a mile to the westward, at the junction of the red and purple Cambrian grits with the Lingula-slates . . . The fry of some large Trilobite first attracted my attention, and then by looking along the ledges, I found fragments (head, body-rings, labrum), but none perfect, of the largest species of *Paradoxides* known, scarcely excepting the great *P. Harlani,* from near Boston. *Agnostus* accompanied it, as usual, being the smallest as *Paradoxides* is the largest, Trilobite of the Primordial zone. Salter's discovery made Porth-y-Rhaw a classic locality for trilobites, but this fame has brought with it successive streams of geologists and collectors, and it is now difficult to obtain more than small

fragments of *Paradoxides.*

Whilst working with the Geological Survey, Salter described and illustrated many trilobites in special Survey publications called *Decades.* Each of these comprised 10 plates, accompanied by detailed descriptions; Salter produced three *Decades* dealing with trilobites, a large proportion of the specimens illustrated originating from Wales and the Welsh Borderland. After leaving the Geological Survey, Salter was able to devote a good deal of his attention to his largest and most important piece of work. This is a monograph in which he intended to illustrate and describe every form of trilobite known from the British Isles. The first part of the work was published in 1864 by the Palaeontographical Society, which was founded in 1847 by professional and amateur geologists with the intention of producing annual volumes devoted to the illustration and description of British fossils, and the series continues to this day. Unfortunately Salter died in 1869 at the early age of 49, long before his monograph was complete, when he had illustrated less than half of the trilobites known at the time. The beautifully produced plates were largely the work of Salter himself, assisted by A. Gawan. Some 500 specimens are illustrated, and of these some 30% originate from Wales, with a further 20% from the Borderland. These figures stress the importance of this area as a source of much of Salter's material.

Apart from his own collecting, and that of Geological Survey Officers, Salter also depended upon a large number of private collectors as a source of many of the specimens that he used. One of these was David Homfray, a fossil collector from Porthmadog, after whom Salter named his giant trilobite from Porth-y-Rhaw *Paradoxides davidis*. Another important local amateur was Dr. Henry Hicks, a physician from St. Davids who made a major contribution to unravelling the history of the ancient rocks of that area. Salter named another large trilobite, *Paradoxides hicksii*, in his honour.

Whilst these early studies were being carried out in the British Isles, European palaeontologists were beginning to investigate trilobites in similar ways. Such was the fame of Welsh trilobites that they soon came to the notice of some of these workers, and it was the French naturalist Alexandre Brongniart who wrote the first full scientific description and gave a formal latinised name to Lhuyd's *flat-fish* in his *Histoire Naturelle des Crustacés fossiles,* published in 1822.

Trilobites and Nomenclature

By the 1820s, nomenclature of all organisms, fossil and living, had become more or less stabilized, following the rationalization of the binomial system of nomenclature by the famous Swedish naturalist Carl Linnaeus in 1758, in the 10th edition of his *Systema Naturae.* This system employs a genus and a species name; with Salter's *Paradoxides davidis,* *Paradoxides* is the generic and *davidis* the specific name. Further examples are well illustrated among trilobites from Wales. Sedgwick, Murchison and Salter are all honoured by trilobite names – e.g. *Angelina sedgwickii, Neseuretus murchisoni,* and *Salterolithus caractaci.* Names may be derived from geographical locations, as is the case with *Flexicalymene cambrensis* (from Cambria, the Roman name for Wales) or they may reflect characteristics intrinsic to the fossils themselves (e.g. *Eodiscus punctatus* – with a punctate dorsal exoskeleton; *Basilicus tyrannus* which is a large species – 'tyrannical'), or, like *Olenus* they may be derived from mythology. Perhaps the most interesting names applied to trilobites are some of those coined in the early days, which reflect the confusion then existing as to their affinities, as beautifully expressed by J.E. Taylor in 1885: 'How

utterly at sea the majority of naturalists were as to the nature of these singular fossils is indicated by some of their generic names. This statement is borne out by such names as *Paradoxides* referring to 'strange' or 'contrary to expectation'; *Agnostus* meaning 'unknown' or 'obscure', *Asaphus* likewise to 'obscure' or 'baffle', *Calymene* to 'concealed', and *Cryptolithus* to 'hidden' or 'concealed'.'

Of all trilobites from Wales none can have had such a complex and chequered nomenclatorial history as Lhuyd's *flat fish.* Brongniart placed them in *Asaphus,* but in 1843 Georg August Goldfuss, Professor of Zoology and Mineralogy at Bonn transferred them to *Ogygia,* to which genus they were usually referred for much of the ensuing threequarters of a century. *Ogygia* was originally proposed as a trilobite name by Brongniart, but unfortunately he was unaware that it had already been used for a moth by the famous German lepidopterist Hübner only shortly before. In nomenclatorial rules no two animals, fossil or living, can bear the same generic name, and since the moth received the name first, a new one had to be sought for the trilobite. The issue was further compounded by problems of interrelationships between the *flat fish* and related trilobites. Thus, since 1822 they have been variously called *Asaphus debuchii, Asaphus Buchii, Ogygia Buchii, Ogygia buchi, Ogygiocaris buchii* and *Ogygiocarella debuchii.* The last is the currently accepted name.

Fragments, Fakes and Distortion

Problems in understanding and classifying trilobites were made even greater by the often fragmentary nature of the fossils, for trilobites, like their distant cousins the crabs and lobsters, periodically shed their hard exoskeleton in order to accommodate increase in size of the individual. Most trilobite fossils are fragments of discarded exoskeletons – complete specimens generally being rather uncommon.

Fossils were often sold to collectors, either by fellow collectors or by local quarrymen. Complete trilobites have always been particularly desirable, and always commanded a far better price than did fragments. Unscrupulous quarrymen often 'repaired' fragmentary trilobites, and skilfully carved segments between detached heads and tails, or added new heads to headless bodies. The end-products were often quite bizarre, combining parts of quite different trilobites. There appear to be no examples of 'repaired' trilobites from Wales, and all the known British examples originate from Dudley in the West Midlands, from where there are two good specimens in the Museum's collection.

The appearance of trilobites can be much altered

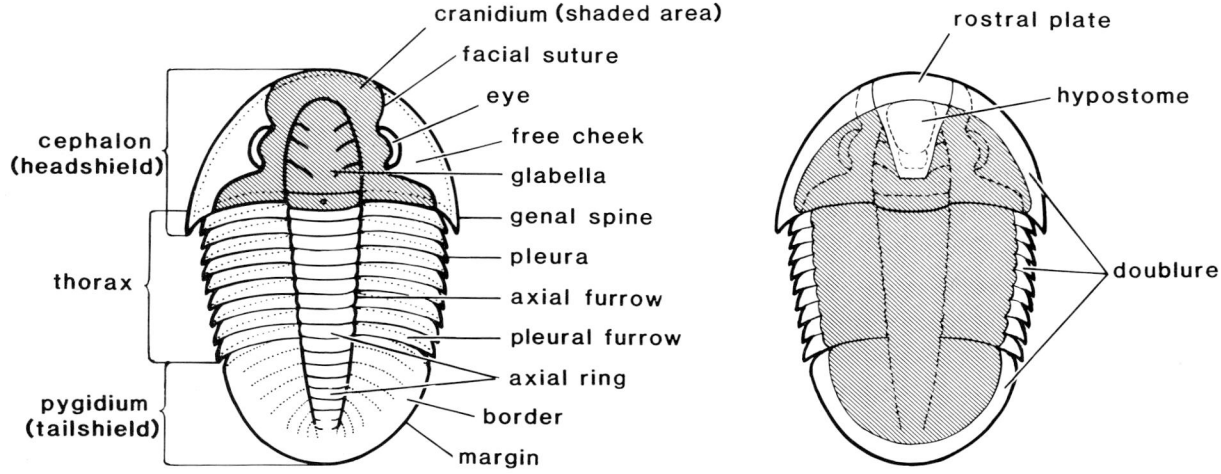

The principal morphological features and terminology of the trilobite exoskeleton. (left) dorsal; (right) ventral.

Three specimens of Angelina sedgwickii, *from the Ordovician (Tremadoc Series) of the Porthmadog district, illustrating the effects of compression from different directions.*

by distortion, and many of the ancient rocks in which they occur have suffered severe folding and compression, as in the case with many of those in north Wales. Often all traces of fossils have been obliterated; the slates near Tremadog, however, are rich in trilobites and *Angelina sedgwickii* is the most distinctive kind from this area. A great variety of distortion is displayed by various examples of this trilobite, some having been compressed from the sides, others from front to back, and others obliquely. The exact proportions of *Angelina sedgwickii* are still somewhat conjectural, as no completely undistorted examples have been found, although there have been numerous attempts to

'straighten out' specimens, most recently by using an elaborate device involving the superimposing of movable grids shown on a television screen, developed by Dr. R.M. Appleby at University College, Cardiff in the 1970s. Distorted specimens have sometimes misled palaeontologists into thinking that more than one species is present at a locality, as happened to Hicks with trilobites from Bay Ogof Hên, Ramsey Island in the last century. His mistake was only discovered by comparison with better-preserved specimens subsequently found elsewhere.

Trilobite appendages and internal organs

It has long been established that trilobites are extinct marine arthropods, a large group of invertebrate animals that includes insects, spiders, crabs and lobsters. Most of the early naturalists appreciated that they were some kind of arthropod; for example, Linnaeus referred to them as *Entomolithus* ('insect-stone') in 1745, and another Swedish naturalist, G. Wahlenberg called them *Entomostracites* ('insect-shells') in 1821. Outside the arthropods, trilobites were placed by Walch in 1771 in the molluscs, and in 1808 the French naturalist P.A. Latreille pronounced that if they had had no legs, they must be Venus molluscs, but if legs were found, then they belonged to the Isopoda (woodlice and their allies) – in the words of H. Wendt (*Before the Deluge*, 1968), 'a truly Solomonic judgement'. The apparent lack of limbs was one of the most perplexing problems for students of trilobites over much of the 19th Century. Only after the discovery of limbs in North America in the latter part of that century could palaeontologists gain a more complete understanding of these fossils. Even today, only a small number are known with remains of limbs, and none have been found in the British Isles. Trilobite appendages are biramous, with a walking leg and an upper branch which probably served as a gill; a recent study by Professor H.B. Whittington of Cambridge University has shown how these limbs might have functioned in the living animal, and this work has implications for the interpretation of certain kinds of tracks and trails commonly found as fossils, and which have been assumed to have been

Meneviella venulosa *(Salter) from the Cambrian of Porth-y-Rhaw, showing branching diverticulae, possibly belonging to the digestive system (x 8). (Photograph courtesy of Mr. M. Lewis).*

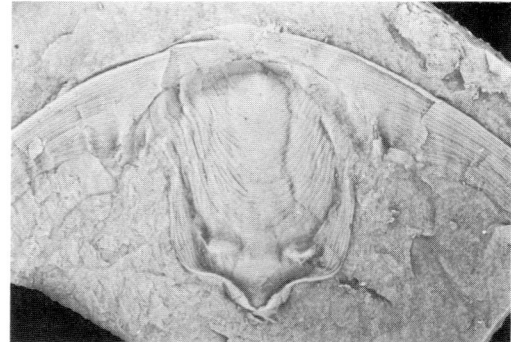

Underside of cephalon of Ogygiocarella debuchii *(Brongniart), showing hypostome in situ (x 2).*

produced by the activity of trilobites. One of the best-known of these is *Cruziana*, a bilobed trail with distinctive diagonal grooves forming a V-pattern, which is very common in the Cambrian rocks in parts of Snowdonia. It had long thought to have been made by a trilobite ploughing through soft mud on the sea bed, but from Whittington's studies it seems unlikely that a trilobite could have generated it. Evidently it was made by some other animal, possibly a soft-bodied one which has left no other trace in the fossil record. Other trails, however, can certainly be ascribed to trilobites, especially some of the ovate, bilobed impressions known as *Rusophycus,* some of which show clearly impressions of genal spines and the tips of segments. These are evidently shallow burrows excavated by trilobites, perhaps made in search of prey, or for concealment. Well preserved examples have been found at Cwm Graianog in Snowdonia.

Remains of internal organs are even rarer than limbs, although there are specimens known with traces of what appear to be musculature and the digestive tract. Some trilobites show structures on the

A trilobite trace-fossil: Rusophycus *from Cwm Graianog, Snowdonia (x 1). (photograph courtesy of Dr. T.P. Crimes).*

lateral parts of the headshield which are possibly branching organs belonging to the digestive system called diverticulae. The Welsh Cambrian species *Meneviella venulosa*, first described by Salter in 1865, shows traces of such diverticulae, and the species name alludes to this character. There is no evidence of a strong jaw apparatus in trilobites, but all bear a rigid plate on the ventral surface of the headshield known as the *hypostome*, which apparently protected the buccal cavity.

Eyes

A most distinctive feature of many trilobites is the pair of prominent eyes on the headshield. When examined with a lens, the trilobite eye can be seen to be compound – made up of a large number of individual facets, reminiscent of the eye in some living arthropods. Trilobite eyes are of special interest on two counts, firstly that they are the most ancient visual system known, secondly in their high degree of sophistication. It has long been recognised that there are two major types. *Holochroal* eyes have contiguous lenses covered by a single, common cornea, and these are found in the vast majority of trilobites. In *schizochroal* eyes each lens is separated from its neighbours, and each has its own cornea.

These eyes are found only in one group of trilobites, the Phacopida, which seem to have evolved from ancestors with holochroal eyes early in the Ordovician Period; the earliest known schizochroal eyes are found in *Ormathops,* a genus that is widespread in rocks of this age in parts of western Dyfed. In many trilobites, including specimens from Wales, the eyes are preserved in calcite. It had long been supposed that this replaced the original eye material during the process of fossilisation, but Dr. E.N.K. Clarkson of Edinburgh University and Dr. K.M. Towe of the Smithsonian Institution, Washington, have both shown that trilobites had calcified eyes in life. A well-known property of calcite is that light rays are split along all crystallographic axes except one, the *c*-axis, along which light will pass as if through glass. Clarkson and Towe discovered that the *c*-axis was normal to the surface in the lenses of the eyes of several different kinds of trilobites with both holochroal and schizochroal eyes. The orientation of the calcite crystals would be expected to be random if the calcite had formed during fossilisation, and it must therefore be assumed that the orientation was original. The implication of this is that trilobites had extremely sophisticated and complex visual systems. A further refinement in schizochroal eyes has been demonstrated by Clarkson and Dr. R. Levi-Setti of Chicago University. They have shown that the two components from which each lens is composed has the bowl-like lower unit separated from the upper part by a wavy surface, which conforms in shape to aplanatic correcting lenses designed by Huygens and

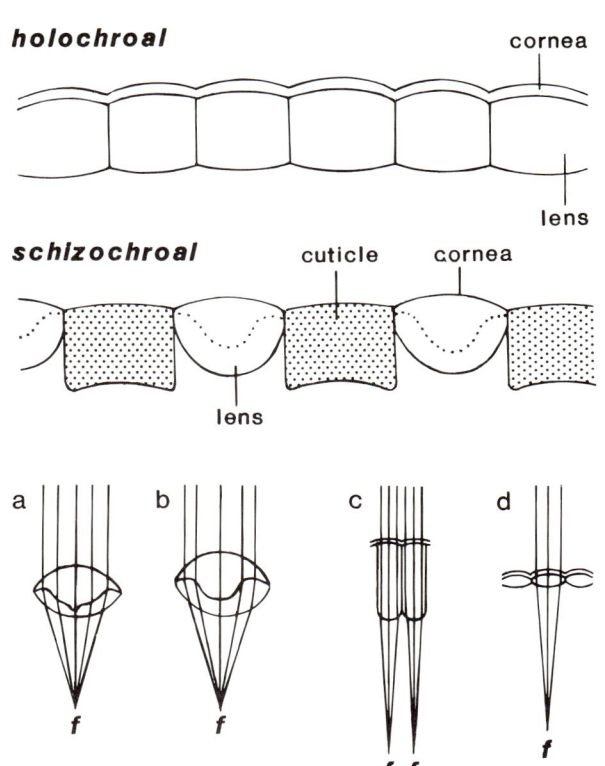

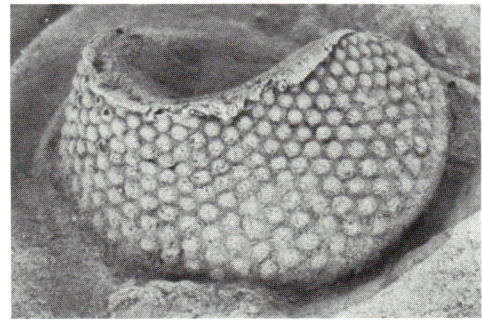

(top left) *Sections through holochroal and schizochroal eyes, showing their contrasting structure.* (left) a, b, *lenses from schizochroal eyes of dalmanitid trilobites conforming to ideal correcting lenses of Des Cartes and Huygens respectively.* c, d, *passage of light through holochroal eyes of an asaphid and olenid trilobite respectively. 'f' indicates where light rays are brought into focus. (After Clarkson, Invertebrate Palaeontology and Evolution, 1979).* (above) *Detail of schizochroal eye of* Dalmanites myops *König, from the Silurian (Wenlock Series) of Usk, Gwent, showing individual lenses, (x 8).*

Des Cartes in the 17th Century. Clarkson and Levi-Setti made experimental models, and the slight differences in the refractive indices between the upper and lower part of the lens operate with the wavy correcting surface to produce a sharp, anastigmatic focus. Trilobites with such sophisticated eyes may have been nocturnal, or perhaps lived in conditions of low illumination.

A great variety of eyes are known and certain groups of trilobites developed enormous eyes. In *Pricyclopyge* they occupied the entire lateral part of the headshield, whilst in *Ellipsotaphrus* and some *Microparia* the eyes merged in the front part of the headshield to produce one vast eye. *Pricyclopyge* and *Ellipsotaphrus* have been found in the lower Ordovician rocks of SW Dyfed, in sediments of comparatively deep water origin. Both are assumed to have been pelagic, a life habit suggested not only by the eyes, but also by the reduced number and narrowness of the thoracic segments.

In contrast to these, there are many trilobites in which the eyes are reduced to only a few lenses, or which were blind. Lack of eyes might be assumed to be a primitive character, but in fact it is a specialised one, for eyeless trilobites undoubtedly evolved from eye-bearing ancestors. Many blind trilobites probably burrowed in soft sediment on the sea bed, or may have lived in conditions of low illumination.

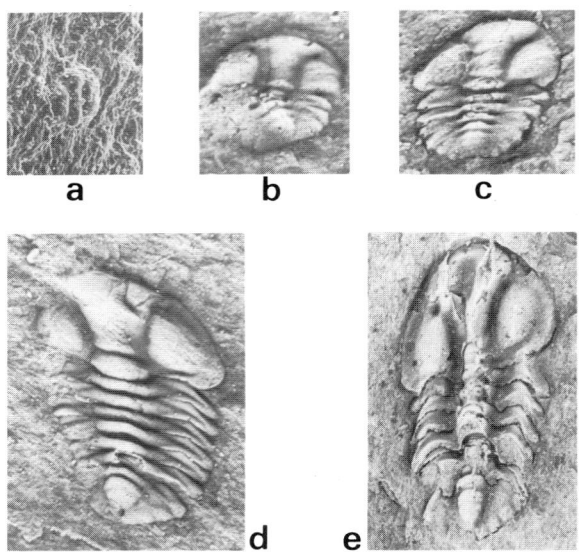

Growth stages of Acontheus, *specimens recently discovered in the Cambrian rocks at Porth-y-Rhaw by Mr. M. Lewis, University College, Cardiff, (a, protaspis (scanning electron microscope photograph), x 20; b, c, meraspides, both x 20; d,e, holaspides, both x 10).*

Trilobites with greatly expanded eyes. **(left)** Cyclopyge grandis *(Salter) with eyes occupying the entire lateral parts of the cephalon (from Ordovician (Arenig Series), Pontyfenni, Dyfed).*
(right) Microparia cf. lusca *Marek, which has one enormous eye, (from Ordovician (Llandeilo Series), Llandrindod Wells, Powys) (Both x 5 approx). (Right hand photograph by courtesy of British Geological Survey, London).*

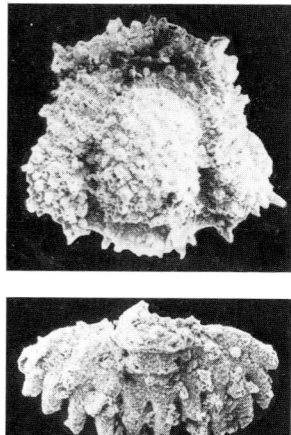

Silicified cephalon and pygidium of Dimeropyge *from the Ordovician (Llandeilo Series) of Cwm Agol, near Dryslwyn Castle, Dyfed. (Scanning electron microscope photographs, both x 20).*

Larvae

Adult trilobites developed from a series of larval stages, the smallest larvae being no larger than a pinhead. Trilobite larvae were first identified and illustrated from Bohemia by the famous French palaeontologist Joachim Barrande. Larval stages have since been found in many parts of the world, and the British palaeontologists Frank Raw (formerly of Birmingham University) and Sir James Stubblefield (formerly Director of the Geological Survey) have illustrated larval suites of certain trilobites found near

Sheinton in Shropshire, and Dr. C.P. Hughes of Cambridge University has illustrated larval stages of *Ogygiocarella* from the Builth district. Because of their small size, trilobite larvae are not easy to find unless one is specifically looking for them, and complete suites are still known only from comparatively few species. However, in certain cases fossils have had the calcium carbonate skeleton replaced by minute quartz crystals, and in this state they are said to be silicified, and can be dissolved out of the enclosing limestone in acid. Silicified trilobites have been found in many areas, including the Llandeilo district, and larval stages occur commonly.

A diversity of forms

Fully grown trilobites range in size from a millimetre to over 50 centimetres, though on average they are in the range of 3 to 4 centimetres long. The rocks of Wales have yielded examples from both extremes of the size range, from the tiny *Shumardia* from lower Ordovician rocks near Arennig Fawr, Gwynedd, to Salter's giant *Paradoxides*. The morphology of many trilobites diverges considerably from the popular conception of these fossils, and those which have been found in Wales include a wide variety of types; those with enormous eyes have already been discussed. A common characteristic of many trilobites is spinosity. Some, like *Cnemidopyge* and *Seleneceme* have greatly elongated spines projecting from the posterolateral corners of the headshield, with another directed straight forwards in front of it. These are both blind forms which probably lived on

the soft mud of the sea bed, and the spines may have acted like skis to prevent the animals from sinking. Yet other trilobites such as the group called odontopleurids possess an array of spines which may have been used as a defence mechanism, whilst the flared spines on the tails of *Cybeloides* and *Deacybele* may have aided in burrowing. These genera had eyes raised on small stalks, like their distant cousins *Encrinurus* and *Atractopyge;* all probably lived partially immersed in sediment on the sea bed, with eyes projecting above the surface. Another trilobite in the same group, *Staurocephalus,* possesses a curious globular expanded lobe on the front of the headshield; its function remains a mystery.

Certain trilobites almost lost the three-lobed form altogether, such as some scutelluids and the homalonotids, in which the thorax consists of broad, strap-like segments with only the faintest indication of where the middle lobe lies. The small, blind agnostids depart even further from the 'typical' trilobite form; they have only two segments in the thorax, and the ovoid head and tail shields look almost exactly the same. It has even been suggested that they are not trilobites at all.

The trinucleids, a group of blind trilobites which includes Lhuyd's *Trinucleum,* are some of the most widespread in Wales. In common with two other groups, the harpedids and dionidids, they bear a prominent flattened border or fringe around the front of the headshield which contains a large number of small pits. In some species these are

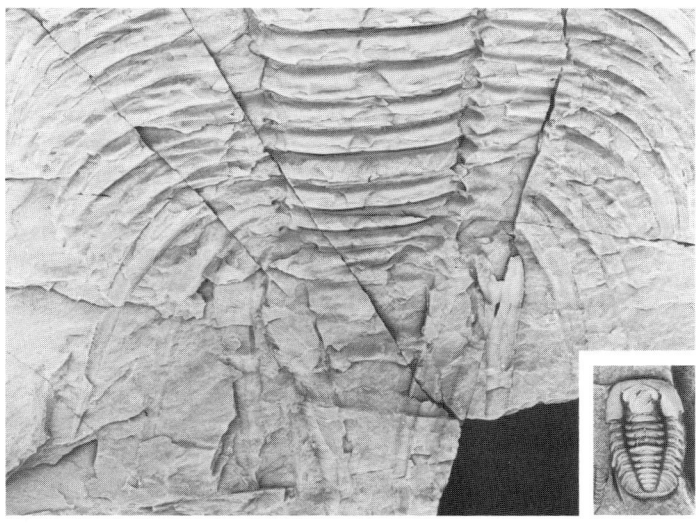

A contrast in size. Part of Paradoxides davidis *Salter, from the Cambrian of Porth-y-Rhaw (x 1) with (inset) adult* Shumardia salopiensis *(Callaway) from the Ordovician (Tremadoc Series) of the Arennig Fawr district (x 5).*

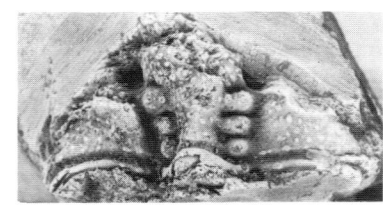

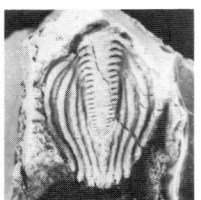

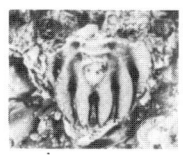

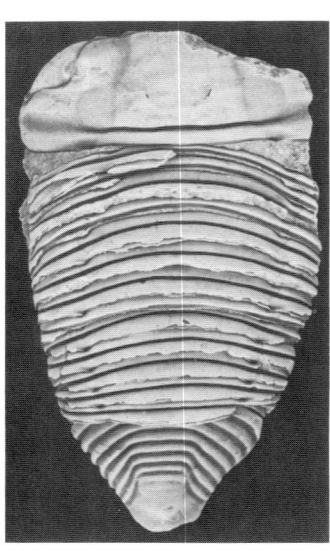

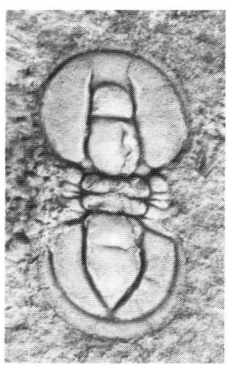

(top left) Seleneceme acuticaudata *(Hicks) from the Ordovician (Llanvirn Series) of the Old Church Stoke district, Powys. Notice the large spines emerging from the cephalon and from the posterior part of the thorax. (x 1.5).* (top centre) *Cranidium and pygidium of* Deacybele pauca *Whittington from the Ordovician (Caradoc Series) of Llanbedrog, Gwynedd. Notice the flared spines on the end of the pygidium (both x 3). (M.V.J. Seaborne Collection).* (top right) *Cranidium (x 5) and pygidium (x 8) of* Staurocephalus clavifrons *Angelin from the Ordovician (Ashgill Series) of the Bala district, Gwynedd. The highly inflated anterior part of the glabella is well shown.* (bottom, far left) *A trilobite with poorly-defined trilobation in the thorax:* Homalonotus knightii *König, from the Silurian (Ludlow Series) of Shobdon, near Presteigne (x 1).* (bottom left) *An 'agnostid' trilobite,* Peronopsis sp. *from the Cambrian (St. David's Series) of Penpleidiau, Caerfai Bay, Dyfed (x 5) (photograph courtesy of Mr. M. Lewis).*

arranged in discrete, ordered rows, while in others the arrangement is more or less random. These pitted fringes comprise two parallel lamellae, and each pit on the upper surface has a counterpart on the lower. In the trinucleids these are joined by a tiny canal, although in harpedids the perforation is proportionally wider. The purpose of the pits has remained a mystery, for they are not found in any extant arthropods. The Australian palaeontologist K.S.W. Campbell suggested that the pitted fringe in trinucleids was a sensory receptor used to indicate changes in current direction – the animal being stable in currents coming from the front, but unstable in other directions; each pit might have had sensory hairs at the base, and the space between the upper and lower layers of the fringe possibly contained digestive glands and an associated circulatory system. In harpedids the broad extended brim might have been used to spread weight like a snowshoe, and in this case the pits may have been used to lighten the body weight, but such elaborate structures probably had other functions as well.

A peculiar feature of many trilobites is that they had the power to roll themselves into a ball, in the fashion of a modern pill bug (woodlouse). Enrollment in trilobites has commonly been interpreted as a defence reaction to afford protection of the soft tissues on the underside of the body. Dr. J. Miller of Edinburgh University, however, has recently offered two further suggestions as to its function. In a swimming trilobite sudden enrollment would result in a drastic reduction in lift, allowing rapid descent, thereby providing effective escape from a predator. Certain trilobites, such as *Placoparia cambriensis* have complex interlocking structures on the undersides of the head and tail shields. The presence of these and the common occurrence of enrolled trilobites in the geological record suggest that they remained tightly enrolled for prolonged periods, perhaps during which anaerobic or other adverse conditions prevailed.

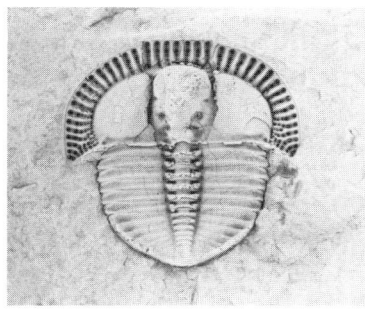

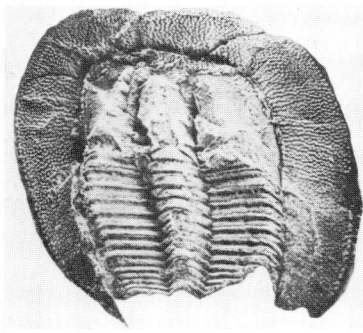

Two trilobites with pitted fringes. **(above)** Trinucleus fimbriatus *Murchison from the Ordovician (Llandeilo Series) of Llandrindod Wells, Powys (x 1.5) (C.T. and I. Taylor collection).* **(below)** Scotoharpes willsi *(Whittington) from the Silurian (Llandovery Series), Bryn-yr-Odin, near Llangollen, Clwyd (x 2).*

An enrolled Calymene blumenbachii *Brongniart from the Silurian (Wenlock Series) of Penylan, Cardiff (x 3).*

Trilobite associations

It has long been realised that particular fossils tend to occur together in assemblages. In the past 25 years these assemblages have attracted a good deal of interest with the development of a branch of palaeontology called palaeoecology, in which attempts are made to identify and describe original communities of animals.

Several examples of trilobite assemblages have been described from Wales in recent years. In the lower Ordovician rocks of the Carmarthen area, Dyfed, Dr. R.A. Fortey of the British Museum (Natural History) and the author have identified three discrete trilobite communities, apparently linked to original water depth, substrate, and prevailing environmental conditions. One of these is of particular interest. It is dominated by olenids, a group seemingly adapted to quiet, oxygen-poor conditions, which Fortey had previously studied in Spitsbergen. These fossils have a large number of wide segments in the thorax, each of which in life would have carried underneath a biramous limb, the upper branch of which was probably a gill. Fortey suggested that a large number of broad gills would increase the effective surface area for oxygen intake, important where oxygen is at a premium. Similar olenid-dominated faunas are well known from upper Cambrian rocks in north Wales.

Specialised deep water associations also occur in the Welsh Ordovician, dominated by large-eyed pelagic trilobites and blind or small-eyed forms. The pelagic trilobites probably lived quite close to the water surface, within the photic zone, whilst the others were benthic, inhabiting areas of low illumination. The fossil assemblage contains a mixture of both associations that settled on the sea bed on death or as moults. Such associations have been found in the lower Ordovician rocks of the Carmarthen-Whitland area, Dyfed, and in the upper Ordovician of the Corris-Aberangell area of southern Gwynedd. In the latter case the deeper water sediments traced north-westwards give way to those of much shallower water origin, and around Bala the Rhiwlas Limestone contains a very diverse and completely different trilobite fauna, from which Professor Whittington described nearly 30 species.

Until recently Silurian trilobites attracted much less attention than have those of the Ordovician, but Dr. A.T. Thomas of Aston University has been studying them in detail over the past decade, and has demonstrated the presence of five different associations in Wales and the Welsh Borderland, ranging from shallow water onshore to deeper water offshore environments. Each association is characterised by the dominance of particular genera, and he has shown that their distribution is related to rock type and supposed position on the continental shelf.

All these examples show that the distribution of trilobites is very much controlled by the original environments in which they lived. In order to infer this, it is not only necessary to study the fossils themselves, but also the total fossil assemblage as well as the sediments in which they occur.

Trilobites in the Welsh geological record

The geological history of trilobites spans some 350 million years. They appeared suddenly early in the Cambrian Period, and some of the oldest Cambrian rocks in many parts of the world contain abundant trilobites. This is not the case, however, in Wales where these rocks have so far yielded very few fossils, even though it is the historical type area for the Cambrian System. This is due in part to the fact that many of these rocks have been strongly deformed, obliterating all traces of fossils that might have been there, and in part to some of them having been deposited in environments which may have been inimical for trilobites and other organisms, or at least those with parts that could be readily fossilised. Even so, in the great thickness of lower Cambrian strata in Wales, at least some fossils might be expected. Their eventual discovery was made in 1887 in the famous Penrhyn Slate Quarries in north Wales, as described by Dr. Henry Woodward in the *Quarterly Journal of the Geological Society of London* in 1888: 'On the 5th August last I received a letter and box of specimens from Professor James J. Dobbie of the University College of North Wales, Bangor, accompanied by the following statement: 'The specimens of Trilobite Nos. 1 and 2, were found by Robert Edward Jones and Robert Lloyd, two quarrymen employed in the Penrhyn Quarry, Bethesda, near Bangor. As no fossils had ever been found in this quarry before, the discovery excited considerable interest in the locality and the quarrymen brought the specimens to the University College, and left them in my hands for examination. Some doubt having been thrown, by residents, on the authenticity of the specimens, I visited the quarry along with the men on the 18th June, and examined them minutely as to the circumstances of the discovery . . . The men showed me the block from where the fossil was taken and I could detect no difference between the slate of which the block is composed and the slate in which the fossil lies imbedded. Whilst searching amongst the debris close by, I found specimen No. 3.' Dr. Woodward, after considering several genera in turn wrote: 'From these considerations I conclude to place the

Penrhyn trilobite in the genus *Conocoryphe,* and I have ventured to dedicate the species to Mrs. Dobbie, under the name of *Conocoryphe Viola'.* Since the publication of Woodward's paper several additional specimens have been found, and some are now in the collections of the National Museum of Wales. They are currently placed in the genus *Pseudatops.*

No other discoveries of trilobites in the early Cambrian of Wales were made until 1958, when Dr. D.A. Bassett found them for the first time in the Hell's Mouth Grits in the cliffs near Abersoch. This discovery also excited considerable interest, and showed that the beds in which they occur are approximately the same age as the Penrhyn Slates, and have helped to contribute towards unravelling some of the complicated Cambrian geology of north Wales, as well as adding important new records of fossils to a sequence of strata which has so far yielded so few.

The oldest abundant trilobite faunas in Wales are found in rocks of Middle Cambrian age (the St. David's Series), where the common appelation for the Cambrian as the 'Age of Trilobites' becomes appropriate so far as the Principality is concerned. These faunas are dominated by the giant *Paradoxides,* which is accompanied by abundant tiny agnostids and eodiscids, as well as a number of other forms. They have been found notably near St. David's, in the Mawddach Valley area near Dolgellau, and south of Maentwrog, and apparently inhabited fairly shallow waters. The *Paradoxides* fauna was replaced by a quite different one in the

Pseudatops viola *(Woodward) from the Cambrian of Penrhyn Quarry, the specimen found by R.E. Jones and R. Lloyd in 1887. (x 1).*

	CAMBRIAN			ORDOVICIAN						SILURIAN				DEVONIAN	CARBONIFEROUS	
	Comley	St. David's	Merioneth	Tremadoc	Arenig	Llanvirn	Llandeilo	Caradoc	Ashgill	Llandovery	Wenlock	Ludlow	Pridoli		Dinantian	Silesian
Eodiscidae	*	*														
Protolenidae	*															
Conocoryphidae	*	*														
Agnostina		*	*	*	*	*	*		*							
Pagetiidae		?														
Paradoxididae		*														
Agraulidae		*														
Solenopleuridae		*														
Dorypygidae		*														
Corynexochidae		*														
Ptychopariidae		*	*	*												
Dikelocephalidae			*													
Parabolinoididae			*													
Hapalopleuridae			*													
Ceratopygidae			*	*												
Olenidae			*	*	*											
Shumardiidae			*	*	*											
Remopleurididae			*	*				*	*							
Asaphidae			*	*	*	*	*	*	*							
Nileidae			*	*	*	*	*									
Dikelokephalinidae				*												
Orometopidae				*												
Cyclopygidae				*	*	*	*		*							
Calymenidae				*	*	*	*	*	*	*	*	*				
Cheiruridae				*	*			*	*	*	*					
Bathyuridae					*											
Bohemillidae					*											
Pliomeridae					*	*										
Alsataspididae					*	*										
Trinucleidae					*	*	*	*	*							
Dionididae					*	*			*							
Illaenidae					*	*		*	*	*						
Raphiophoridae					*	*	*	*	*		*	*				
Odontopleuridae					*	*	*	*	*	*	*	*				
Encrinuridae					*	*	*	*	*	*	*	*				
Dalmanitidae					*	*			*	*	*	*				
Homalonotitidae						*		*	*	*	*	*				
Lichidae						*	*	*	*	*	*	*				
Dimeropygidae						*										
Scutelluidae						*			*	*	*					
Proetidae						*	*	*	*	*	*	*			*	*
Aulacopleuridae						*	*	*	*	*	*	*			*	
Calmoniidae							*	*	*	*	*	*	*			
Pterygometopidae							*	*								
Phillipsinellidae							*	*								
Harpedidae							*	*		*	*					
Staurocephalidae									*		*					
Phacopidae										*	*					
Brachymetopidae											*	*			*	*

Stratigraphical distribution of different families of trilobites in Wales. N.B. The Agnostina includes several families of 'agnostid' trilobites.

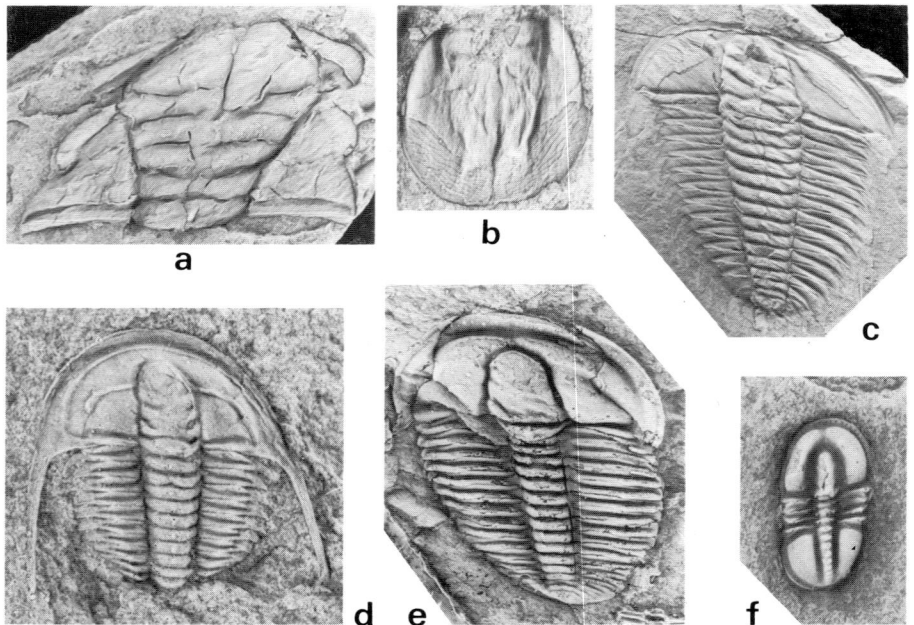

CAMBRIAN TRILOBITES
a, b, *Cranidium and pygidium of* Paradoxides hicksii *Salter, St. David's Series, Nine Wells, St. Davids (both x 2);* **c,** Parabolina spinulosa *(Wahlenberg), Merioneth Series, Nant-y-Gist-faen, near Arennig Fawr, Gwynedd (x 1.5);* **d,** Hamatolenus (Myopsolenus) douglasi *Bassett, Owens & Rushton, Comley Series, E side of Hell's Mouth, Gwynedd (x 2);* **e,** Parasolenopleura applanata *(Salter),* St. Davids Series, Porth-y-Rhaw, St. Davids (x 3.5) (M. Lewis collection); **f,** Eodiscus punctatus *(Salter), horizon and locality as* **a** *(x 7).*

Upper Cambrian (Merioneth Series), which was of low diversity and dominated by olenids, which as suggested above, probably flourished in quiet, poorly oxygenated seas. In Cambrian strata trilobites have been successfully used as zone fossils, the zonal indices being principally paradoxidids and agnostids in the Middle and olenids in the Upper Cambrian. Many of these zones were established in Scandinavia where there are trilobite faunas similar to those in Wales. It was not until recently, however, that the topmost zone of the Cambrian was identified for the first time in Wales by Dr. A.W.A. Rushton of the British Geological Survey. It contains an interesting assemblage of trilobites including the earliest representatives in Wales of families such as the Asaphidae, Nileidae and Shumardiidae which become important in the Ordovician.

The Ordovician Period saw the acme of the Trilobita throughout the world, when the greatest diversity of genera, species and morphology was achieved. Ordovician rocks in Wales were deposited in a wide variety of environments, and as a consequence they contain particularly diverse fossil faunas. The oldest Ordovician rocks, the Tremadoc Series, contain trilobites of both 'Cambrian' and 'Ordovician' aspect, and for this and other reasons

there has long been controversy surrounding the classification of the Tremadoc, but now the majority of geologists place it in the Ordovician System. Its faunas in Wales are dominated by the groups that appeared at the end of the Cambrian noted above, as well as including olenids such as the well-known *Angelina sedgwickii* and the earliest representatives of the most important of the large-eyed groups in Wales, the cyclopygids. Whilst conditions in the Tremadoc were fairly uniform with the deposition of mudstones over much of Wales, a much greater variety of environments are represented by the succeeding Arenig and Llanvirn strata, which range from shallow-water inshore to deep offshore deposits, the latter containing the fauna of large-eyed and blind trilobites described previously. Many trilobite families make their first appearance in Wales at this time, including the trinucleids which subsequently form such an important part of Welsh Ordovician trilobite faunas.

Sediments deposited in the middle of the Ordovician Period, comprising the Llandeilo Series, have yielded some of the most abundant trilobites in Wales, and are probably the best known to collectors; it is these strata that provided many of Lhuyd's specimens from the Llandeilo area, and in

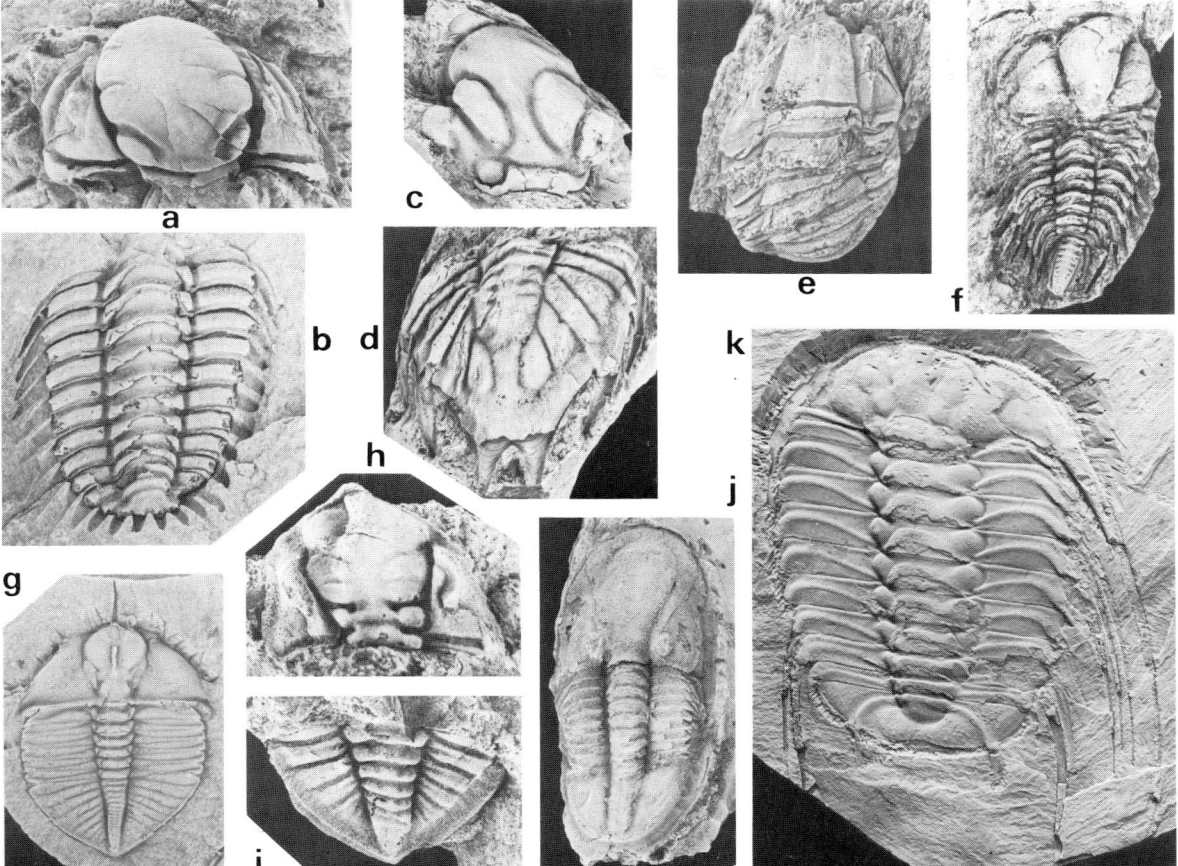

ORDOVICIAN TRILOBITES

a, b, *Cephalon, and thorax and pygidium of* Pseudosphaerexochus octolobatus *(M'Coy), Ashgill Series, Moel Fferna, Cynwyd, Clwyd (both x 2);* **c, d,** *cranidium and pygidium of* Platylichas nodulosus *(M'Coy), Caradoc Series, Bala district, Gwynedd (both x 2);* **e,** *cephalon and thorax of* Brongniartella minor *(Salter), horizon and locality as* **c** *(x 2);* **f,** Dindymene longicaudata *Kielan, Ashgill Series, Llanbedrog district, Gwynedd (x 5);* **g,** Cnemidopyge bisecta *(Elles), Llandeilo Series, Llandrindod Wells, Powys (x 1.5);* **h, i,** *cephalon and pygidium of* Kloucekia apiculata *(M'Coy), horizon and locality as* **c** *(both x 2);* **j,** Stygina *cf.* latifrons *(Portlock), horizon and locality as* **a** *(x 1.5);* **k,** Selenopeltis inermis *(Klouček), Llanvirn Series, Llanvirn, Dyfed (x 1). (**a, b,** G. Thomson collection;* **f,** M.V.J. Seaborne collection;* **g,** C.T. and I. Taylor collection).*

the Builth-Llandrindod area localities such as Llanfawr Quarry are particularly rich and well-known sources. Besides trinucleids and asaphids (the latter including *Ogygiocarella debuchii*) numerous specimens of the long-spined *Cnemidopyge* occur. The succeeding Caradoc faunas, which include those found in 'Trilobite Dingle', are dominated by a succession of trinucleid species, calymenids, homalonotids, and calmoniids, all found in sediments of comparatively shallow water origin, and particularly rich faunas have been found in the Bala and Pwllheli districts. The fossils from the summit of Snowdon are of Caradoc age, and although dominated by brachiopods do include rare trilobites (mostly trinucleids). The richest and most diverse Ordovician trilobite faunas, however, are found in the Ashgill, and shallow water deposits such as the Rhiwlas Limestone near Bala and the Dolhîr Formation near Cynwyd have yielded representatives of over 20 different families. One of Lhuyd's trilobites, *Atractopyge verrucosa* originated from Ashgill strata near Llandeilo. Deeper water sediments of Ashgill age with large-eyed trilobites have been mentioned earlier.

The close of the Ordovician Period saw the demise of a large number of trilobites, and many important families, including the trinucleids, asaphids and cyclopygids became extinct. A contributory factor to this extinction may have been a general cooling of the oceans brought about by an ice-age that occurred at the end of the period.

The lowest Silurian rocks in Wales (Llandovery Series) are not rich in trilobites, and where they do occur (for example in fairly shallow-water sediments near Haverfordwest, Llandovery and Meifod) are dominated by calymenids, phacopids and encrinurids. Deeper water, offshore sediments have yielded a few trilobites near Llanystumdwy, but in contrast to similar Ordovician deposits there are no large-eyed pelagic forms, and after the Ordovician this niche must have been occupied by animals other than trilobites. Wenlock rocks contain far more abundant and diverse trilobites, and Penylan Quarry, Cardiff was long renowned as a rich source which yielded many complete specimens, in particular calymenids, encrinurids, phacopids and dalmanitids. Unfortunately this locality has now largely disappeared under a road. Other areas have strata of this age deposited in various environments as described by Thomas. For example, shallow water limestones such as those near Usk have a proetid-dominated fauna, whilst mudstones cropping out in the same area are dominated by *Dalmanites*. A different kind of limestone exposed near Old Radnor, which was deposited on a topographical 'high' in a more offshore position than that of Usk, contains trilobites resembling those found in similar environments in Bohemia, but unlike many of their British contemporaries. In south Pembrokeshire there are shallow-water onshore sandstones with a very limited trilobite fauna, containing only calmoniids and homalonotids. Sediments of deeper water origin that crop out in central and north Wales generally contain few trilobites, although they are locally abundant. Near Builth Wells Professor O.T. Jones coined the term '*Acidaspis* Limestone' for a deposit containing the abundant remains of an odontopleurid trilobite, whilst near Llanrwst one horizon has yielded a specialised fauna with a dalmanitid and odontopleurid which is particularly widespread, occurring also in the Long Mountain near Welshpool, in the Howgill Fells, Cumbria, and as far afield as Scania, southern Sweden.

The Ludlow Series in Wales sees a progressive decline in the numbers and diversity of trilobites concomitant with a steady shallowing of the sea. Those in the earlier part of the series are much like those of the Wenlock, but younger sediments contain different species of calymenids, encrinurids, dalmanitids, lichids, and proetids. In some of these rocks trilobites can be locally abundant, and in 1978

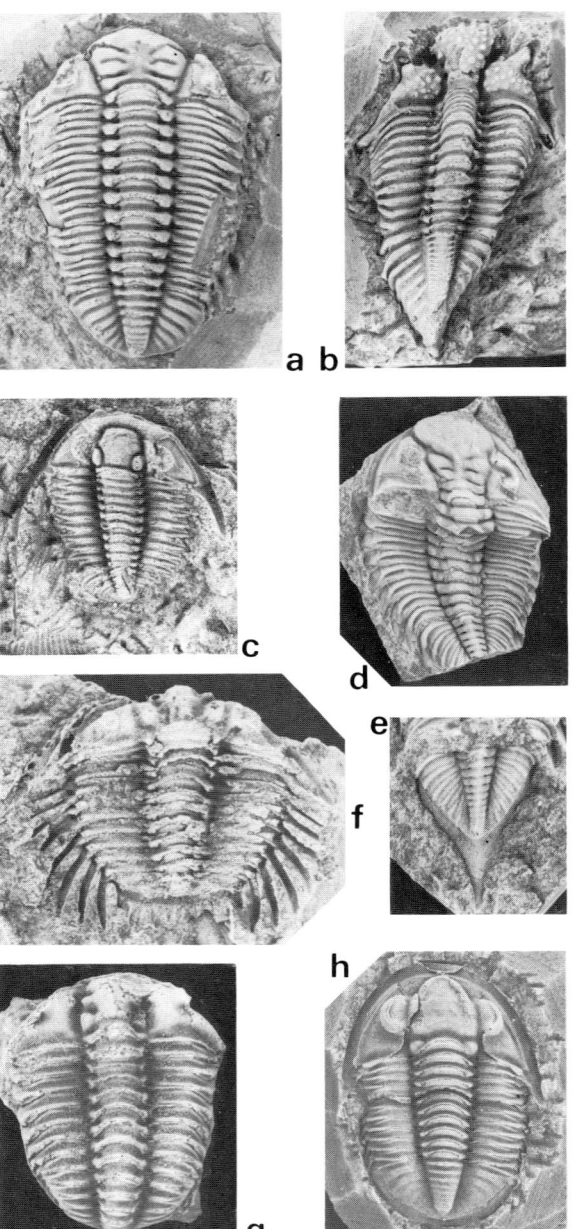

SILURIAN TRILOBITES
a, Acastella spinosa *(Salter), Ludlow Series, Llandegfedd, Usk, Gwent (x 2);* b, Encrinurus tuberculatus *(Buckland), Wenlock Series, Penylan, Cardiff (x 1.5);* c, Harpidella *sp., horizon and locality as* a *(x 3);* d, e, *cephalon and thorax, and pygidium of* Dalmanites myops *(König), horizon and locality as* b *(x 1.5 and x 2 respectively);* f, Leonaspis coronata *(Salter), horizon and locality as* b *(x 3);* g, Calymene puellaris *Reed, horizon and locality as* a *(x 2.5);* h Proetus obconicus *Lindström, horizon and locality as* a *(x 2.5). (*a, c, g, h, *C.T. and I. Taylor collection).*

CARBONIFEROUS TRILOBITES
a, Cummingella *aff. carringtonensis (Woodward), Dinantian, near Linney Head, Castlemartin, Dyfed (x 2)*; b, c, *cephalon and pygidium of* Brachymetopus *sp., Silesian, Coal Measures, Cefn Coed Marine Band, near Tondu, Mid Glamorgan (both x 20)*; d, Paladin *sp., Dinantian, Trefor Rocks, Llangollen, Clwyd (x 3).*

Mr. C.T. and Mrs. I. Taylor of Barry found over a hundred specimens of *Calymene*, many of which were complete, in a temporary trench near Llandegfedd Reservoir, Usk. The uppermost Ludlow strata are characterised by a homalonotid and a calmoniid, a similar shallow water association to that found in the Wenlock in south Pembrokeshire. The sediments of the succeeding Přídolí Series in Wales are mostly in non-marine Old Red Sandstone facies, although a calmoniid has been found at its base in the Cennen Valley near Llandeilo. No trilobites have been found in the Devonian of Wales, since nearly all deposits of that age in the area are of non-marine origin.

Trilobites reappear in the Welsh geological record in the Carboniferous with the return of marine deposits, but only three families are left. They have been found in the Carboniferous Limestone in SW Dyfed, on the Gower, and in the Merthyr Tydfil and Llangollen districts among others, but are most abundant where carbonate mound or 'reef' deposits occur, such as in the Llandudno and Prestatyn areas. The last known trilobites are found in rocks of Permian age, but in Britain conditions became unsuitable for trilobites some time before that, when the great Coal Measures swamps were in existence. Sometimes the sea engulfed these swamps temporarily, and one of these marine incursions brought with it the last known British trilobites. These tiny creatures, barely a centimetre long, have been found in SW Dyfed and in Mid Glamorgan. Thus, besides some of the earliest specimens, Wales also has some of the last trilobites in Britain.

Acknowledgements
I am grateful to Dr. T.P. Crimes, Mr. M. Lewis, Mr. S.F. Morris, Dr. W.H.C. Ramsbottom, Professor B.F. Roberts, Dr. A.W.A. Rushton, Mr. J.Thackray and Mr. C.T. and Mrs. I. Taylor for the supply or loan of specimens or of illustrative material. Dr. E.N.K. Clarkson and Dr. P.D. Lane kindly read the manuscript and offered suggestions for its improvement.

Further reading
The literature on trilobites is extensive. A list of the more general works is given in:
CLARKSON, E.N.K. 1979. *Invertebrate Palaeontology and Evolution.* x + 323 pp., George Allen & Unwin, London, Boston and Sydney.

The following monographs of the Palaeontographical Society deal specifically with, or include Welsh trilobites.
LAKE, P. 1906-46. A monograph of the Cambrian trilobites, 350 pp., 47 pls.
LANE, P.D. 1971. British Cheiruridae (Trilobita), 95 pp., 16pls.
OWENS, R.M. 1973. British Ordovician and Silurian Proetidae (Trilobita), 98 pp., 15 pls.
SALTER, J.W. 1864-83. A monograph of the British trilobites from the Cambrian, Silurian and Devonian formations, 224 pp., 30 pls.
TEMPLE, J.T. 1970. The Lower Llandovery brachiopods and trilobites from Ffridd Mathrafal, near Meifod, Montgomeryshire, 76 pp., 19 pls.
THOMAS, A.T. 1978-81. British Wenlock trilobites, 100 pp., 25 pls. (continuing).
WHITTARD, W.F. 1955-67. The Ordovician trilobites of the Shelve Inlier, west Shropshire, 352 pp., 50 pls.
WHITTINGTON, H.B. 1950. British trilobites of the family Harpidae, 55 pp., 7 pls.
WHITTINGTON, H.B. 1962-68. A monograph of the Ordovician trilobites of the Bala area, Merioneth. 138 pp., 32 pls.
WOODWARD, H. 1883-84. A monograph of the British Carboniferous trilobites, 86 pp., 10 pls.

Lists and range charts of all described British trilobite species, together with a comprehensive bibliography are given in:
THOMAS, A.T., OWENS, R.M. & RUSHTON, A.W.A. 1984. Trilobites in British stratigraphy. *Geological Society of London Special Report* no.16, 78 pp.

GEOLOGICAL TIME SCALE

ERA		PERIOD	AGE (Millions of years ago)	Summary of geological history of Wales
CENOZOIC	QUATERNARY	Recent or Holocene	0.01 – present	Coastal areas drowned by rising sea level caused by melting ice at the end of the Ice Age. Deposits of alluvium and peat, with further development of present drainage patterns and modern flora and fauna.
		Pleistocene	1.6 – 0.01	The 'Ice Age', with repeated glaciations and milder interglacial periods. The most recent major glaciation reached a maximum 18,000 years ago, and the last local ice left North Wales by 14,500 years ago. Modification of land forms by ice scouring and deposition of glacial drift. First evidence of Man in Wales from Pontnewydd Cave, North Wales, about 200,000 years old.
	TERTIARY	Neogene	26 – 1.6	Prolonged, pulsatory uplift and erosion. Late Palaeogene and early Neogene terrestrial sediments known only from Mochras and locally in Gwynedd and south-west Dyfed. Early Palaeogene intrusive igneous rocks in north-west Wales. Basic landforms and drainage patterns established.
		Palaeogene	65 – 26	
MESOZOIC		Cretaceous	140 – 65	No rocks of this age known in Wales. Different opinions suggest either a persistence of terrestrial conditions, or that the Chalk sea covered much of the country.
		Jurassic	195 – 140	Lower Jurassic marine rocks known only from south-east Glamorgan and Mochras borehole, Harlech, with thick, younger sediments in Bristol Channel. Warm, shallow sea may have transgressed over Wales through the period but direct evidence is lacking.
		Triassic	230 – 195	Arid and semi-arid terrestrial conditions, with evidence of periodic flash floods. Dinosaur footprints and early mammal remains known from Glamorgan. Marine transgression in south-east Wales at the very end of the period.
PALAEOZOIC	UPPER	Permian	280 – 230	Uplift and mountain building (Hercynian Orogeny). Erosion across most of Wales.
		Carboniferous	345 – 280	Marine transgression early in period, leading to spread of warm, subtropical seas with rich coral/brachiopod faunas; extensive deposition of carbonate sediments that now form the Carboniferous Limestone. Regression in middle of period, with widespread deltaic deposits. Rich vegetation on coastal plains and deltas in late Carboniferous times; peat accumulated to form coal seams of the Coal Measures.
		Devonian	395 – 345	Uplift and mountain building (Caledonian Orogeny) continued from Silurian times, resulting in the deposition of the terrestrial Old Red Sandstone across most of Wales. Rapid diversification of land floras and non-marine fish faunas.
	LOWER	Silurian	435 – 395	Marine muds, silts and sands, with local carbonate sediments. Land with deltas across South Wales. Shallowing and retreat of the sea late in the period, with widespread onset of terrestrial conditions. Earliest fishes in Wales, and first vascular land plants appeared.
		Ordovician	505 – 435	Marine muds, sands, grits and local carbonate sediments. Evidence of extensive volcanicity in north, mid and south-west Wales. Fossil faunas increasingly diverse, including first corals.
		Cambrian	570 – 505	Transgression of sea into Wales, with deposition of grits, sands and muds. First abundant fossils, including earliest trilobites and brachiopods.
PRECAMBRIAN			4600 – 570	Oldest rocks in Wales dated at about 600-700 million years old, but are possibly considerably older. Evidence of intermittent marine conditions and volcanicity, with periods of folding, uplift and erosion. Earliest fossils from Wales are 'jellyfish' from Carmarthen of late Precambrian age.